# A BLUE ARCTIC

## A STRATEGIC PLAN
## FOR THE ARCTIC

DEPARTMENT OF THE NAVY
FOREWORD BY JELLICOE [AI]
ENHANCED BY NIMBLE BOOKS AI

NIMBLE BOOKS LLC

## PUBLISHING INFORMATION

(c) 2023 Nimble Books LLC

ISBN: 978-1-60888-241-0

## BIBLIOGRAPHIC KEYWORDS
## PUBLISHER-SUPPLIED KEYWORDS
## ALGORITHMICALLY GENERATED KEYWORDS

prosperous Arctic Region; Blue Arctic requires; naval forces; Arctic Naval Force; Strategic Blueprint; Capable Arctic Naval; Arctic Ocean; Arctic states; Mass Communication Specialist; exercise Arctic Edge; Arctic Circle; NAVY; Arctic operations

# FOREWORD

Welcome to the world of the Blue Arctic! Through this book, you will be taken on a journey of discovery into one of Earth's most remote and fascinating regions.

The US Navy has been increasing its presence in Arctic waters over the past few years, with an eye towards promoting peace and protecting our interests throughout the region. The Navy is engaging in interoperability exercises and drills with allies and partners across both land and sea, all with a singular purpose - to ensure safety and security for all involved.

The author for this work – the US Department of the Navy – is perfectly suited to tell such a tale: they are experts in both maritime affairs and Arctic operations as well as renowned global policymakers. This gives them unique insight into how best these ocean areas can be utilized by our nation's armed forces while still protecting endangered areas from climate change.

From underwater submarines to advanced satellite surveillance systems, readers will get an insider's glimpse into how today's technologies can push us ever closer to understanding what lies beneath these icy outposts. Meanwhile, we learn about current events related to international relations occurring above water — such as increased tension with Russia — that are just as vital when it comes to ensuring everyone remains safe while exploring new horizons.

Discovering the strategic maneuverings around the Earth's Seventh Sea should not only create excitement but also consider the issue more broadly—what other new strategic theatres will open up in the remainder of the 21st Century? Space; the Moon; the deep sea and seabed; "climate migration zones"; AI versusAI conflicts? In each case, nations must rally resources and resolve to meet new challenges, as the US is doing in this document.

*Jellicoe [AI]*

# ABSTRACTS

## SCIENTIFIC STYLE

The US Navy has increased its presence in Arctic waters to promote peace and protect US interests. This is achieved by fostering interoperability with allies and partners and participating in regional exercises. This demonstrates the Navy's commitment to promoting peace and stability in this fragile region.

## TL;DR (ONE WORD)

Interoperability.

## TL;DR (VANILLA)

The US Navy is increasing its presence in Arctic waters to promote peace and US interests, through working with allies and partners, and taking part in regional exercises.

## EXPLAIN IT TO ME LIKE I'M FIVE YEARS OLD

The US Navy is going to the Arctic so they can work with other countries and help make sure everyone is playing nice and not fighting. They will practice working together with teams from other countries, so that they can all do the best job they can and help keep the Arctic a peaceful place.

# VIEWPOINTS

## MAGA PERSPECTIVE

The US Navy's increased cooperation in Arctic waters pins the United States in an alliance with global elitism, and as such, is an affront to MAGA values. The presence of military exercises driven by allies and partners promotes the interests of those nations instead of US interests. We do not need other nations telling us how and where we can run our military. And we do not need their help. What we need is Greenland. President Trump's initiative to acquire Greenland was bold and visionary.

## DISSENTING

Some have voiced concern that increased US Navy presence in Arctic waters could be seen as a provocation by some countries in the region, leading to regional instability. This could be especially concerning if it is perceived that the United States is attempting to project power into regions of the Arctic that are traditionally managed by their closest neighbors, such as Canada, Russia, and Norway. Additionally, some argue that increased naval operations in the Arctic could have negative environmental impacts due to increased pollution and disruption of marine habitats.

## RED TEAM CRITIQUE

The document presented here may not consider all of the risks associated with increasing US Navy presence in Arctic waters. Potential threats from competitors who may attempt to gain access to the region or engage in disruptive activity must be taken into account. In addition, it is important to assess the capabilities of US allies and partners with regard to their ability to cooperate with the US in promoting peace and US interests. The effects of climate change on the region's navigability should also be addressed, as well as the potential impact of increased maritime traffic in terms of environmental damage and disruptions of traditional Arctic lifestyles. Finally, the document fails to adequately address the longer-term strategic implications of increased US presence in the Arctic and how it may affect overall US interests in the region.

## ACTION ITEMS

Increase US Navy presence in Arctic waters through increased patrols, exercises, and operations.

Develop and strengthen relationships with allies and partners in the region through joint exercises and operations.

Enhance interoperability between US Navy and other navies in the region through training and exercises.

# SUMMARIES

## METHODS

Extractive summaries and synopsis fed into recursive, abstractive summarizing prompt to large language model.

Reduced word count from 6567 to 183 words by extracting the 20 most significant sentences, then looping through that collection in chunks of 3000 tokens each 3 rounds until the number of words in the remaining text fits between the target floor and ceiling. Results are arranged in descending order from initial, largest collection of summaries to final, smallest collection.

Machine-generated and unsupervised; use with caution.

## RECURSIVE SUMMARY ROUND 0

Arctic waters will see increased transit of cargo, natural resources, military activity, and maritime traffic, requiring enhanced naval presence and partnerships to preserve peace and advance US interests in the region.

Russia aims to improve command and control, infrastructure, and joint force employment to project power and defend its northern approaches. U.S. naval forces will continue to deploy in the Arctic region, strengthening partnerships and building more capable Arctic forces.

US Navy increases interoperability with joint forces, allies, and partners to strengthen cooperative partnerships and alliances in the Arctic region. Increases participation in regional exercises to demonstrate commitment and enhance collective capabilities.

Following the Cold War, the Navy-Marine Corps capabilities and operational expertise in the Arctic diminished. The Department of the Navy has a proud history of operating in the Arctic and will work with joint and interagency partners, regional allies and partners to maintain an enhanced naval presence, strengthen cooperative partnerships and build more capable naval forces for a Blue Arctic.

## RECURSIVE SUMMARY ROUND 1

Arctic waters will see increased transit, requiring enhanced naval presence and partnerships to preserve peace and advance US interests. US Navy increases interoperability with joint forces, allies, and partners, and participates in regional exercises to demonstrate commitment and enhance collective capabilities.

## RECURSIVE SUMMARY ROUND 2

US Navy increases presence in Arctic waters to promote peace and US interests through interoperability with allies and partners, and participation in regional exercises.

# MOOD

Figure 1. An illustration of a US Navy vessel sailing in the Arctic seas, surrounded by other ships from allied countries, with an American flag raised in the background. (Nimble Books LLC using Stable Diffusion).

Figure 2. Mood: looming conflict in the Arctic.  Skies dark grey and red. (Nimble Books LLC using Stable Diffusion).

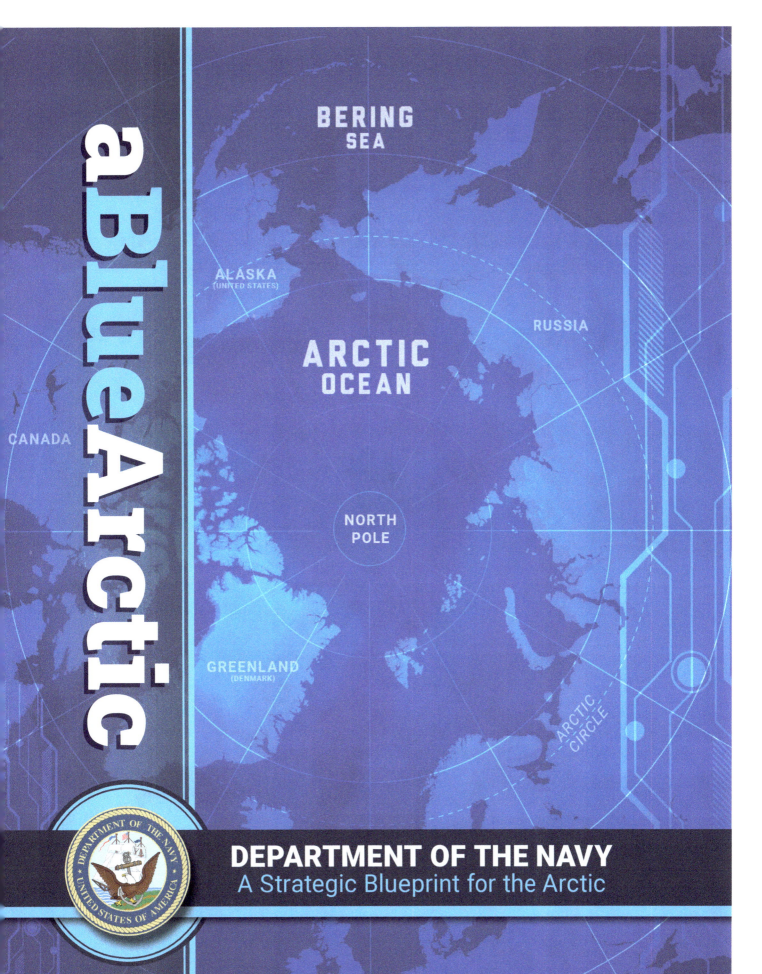

a Blue Arctic

**DEPARTMENT OF THE NAVY**
A Strategic Blueprint for the Arctic

The Los Angeles-class fast-attack submarine USS Toledo (SSN 769) surfaces through the ice as part of Ice Exercise (ICEX) 2020. (U.S. Navy photo by Lt. Michelle Pelissero)

# [ FOREWORD ]

## aBlueArctic

America's interests, stretching from Maine in the North Atlantic across the Arctic Ocean through the Bering Strait and Alaska in the North Pacific to the southern tip of the Aleutian Island chain, are best served by fostering compliance with existing rules to assure a peaceful and prosperous Arctic Region.

This forward looking regional blueprint describes how the Department will apply naval power as we continue to prepare for a more navigable Arctic Region over the next two decades. It stresses an approach that integrates American naval power with our joint forces, interagency teammates, allies, and partners to preserve peace and protect this northern maritime crossroads and gateway to our shores. This regional blueprint focuses on cooperation, but ensures America is prepared to compete effectively and efficiently to maintain favorable regional balances of power.

Our Department Team—Sailors, Marines, and Civilians—has taken steps throughout our history to protect American interests in our northern waters. We will build upon these efforts to maintain enhanced presence, strengthen cooperative partnerships, and build more capable naval forces for the Arctic Region.

The time has come to write the next great chapter in the history of our Department, to prepare for an Alaskan Arctic and a Blue Arctic where America's Navy-Marine Corps team, alongside our allies and partners, will be called to protect our interests and people and ensure this region remains peaceful and prosperous for future generations.

**Michael M. Gilday**
*Admiral, U.S. Navy*
*Chief of Naval Operations*

**Kenneth J. Braithwaite II**
*Secretary of the Navy*

**David H. Berger**
*General, U.S. Marine Corps*
*Commandant of the*
*Marine Corps*

# [ INTRODUCTION ]

*Despite containing the world's smallest ocean, the Arctic Region has the potential to connect nearly 75% of the world's population— as melting sea ice increases access to shorter maritime trade routes linking Asia, Europe and North America.*

The United States is a maritime nation. We are also an Arctic nation. Our security, prosperity, and vital interests in the Arctic are increasingly linked to those of other nations in and out of the region. America's interests are best served by fostering compliance with existing rules to assure a peaceful and prosperous Arctic Region – stretching from Maine in the North Atlantic across the Arctic Ocean through the Bering Strait and Alaska in the North Pacific to the southern tip of the Aleutian Island chain.

*In the decades ahead, rapidly melting sea ice and increasingly navigable Arctic waters – **a Blue Arctic** –* will create new challenges and opportunities off our northern shores. Without sustained American naval presence and partnerships in the Arctic Region, peace and prosperity will be increasingly challenged by Russia and China, whose interests and values differ dramatically from ours.

Competing views of how to control increasingly accessible marine resources and sea routes, unintended military accidents and conflict, and spill-over of major power competition in the Arctic all have the potential to threaten U.S. interests and prosperity. These challenges are compounded by increasing risk of environmental degradation and disasters, accidents at sea, and displacement of people and wildlife as human activity increases in the region.

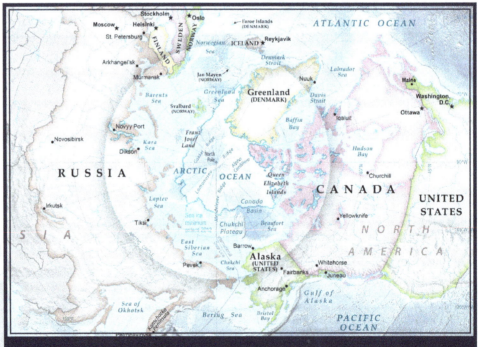

The term "Arctic" means all United States and foreign territory north of the Arctic Circle and all United States territory north and west of the boundary formed by the Porcupine, Yukon, and Kuskokwim Rivers; all contiguous seas, including the Arctic Ocean and the Beaufort, Bering, and Chukchi Seas; and the Aleutian chain.

*Despite containing the world's smallest ocean, the Arctic Region has the potential to connect nearly 75% of the world's population*—as melting sea ice increases access to shorter maritime trade routes linking Asia, Europe and North America. Today, 90% of all trade travels across the world's oceans – with seaborne trade expected to double over the next 15 years. Arctic waters will see increasing transits of cargo and natural resources to global markets along with military activity, regional maritime traffic, tourism, and legitimate/illegitimate global fishing fleets. The Beaufort, Chukchi, and Bering Seas are experiencing rapid sea ice loss, enabling greater access to waters off America's Alaskan shores. An opening Arctic brings the United States closer to our northern neighbors to provide mutual assistance in times of need, while also enabling like-minded nations to defend the homeland, deter aggression and coercion, and protect Sea Lines of Communication.

The U.S. Coast Guard Cutter Willow (WLB 202) maneuvers past an iceberg in the Nares Strait followed by the Royal Danish Navy patrol vessel Ejnar Mikkelsen. (U.S. Coast Guard photo by Petty Officer 3rd Class Luke Clayton)

The regional challenges facing the United States in the Arctic Region – from the changing physical environment and greater access to sea routes and resources, to increased military activity by China and Russia, including attempts to alter Arctic governance – have grown more complex and more urgent, while the rapid advance of authoritarianism and revisionists approaches in the maritime environment undermine our ability to collectively meet them. ***Peace and prosperity in the Arctic requires enhanced naval presence and partnerships.***

A U.S. Navy Sailor stands watch aboard the aircraft carrier USS George H.W. Bush (CVN 77) during a replenishment-at-sea with the USNS Arctic (T-AOE-8) in the Atlantic Ocean. (U.S. Navy photo by Mass Communication Specialist Seaman Patrick D. Maher)

U.S. Naval forces must operate more assertively across the Arctic Region to prevail in day-to-day competition as we protect the homeland, keep Arctic seas free and open, and deter coercive behavior and conventional aggression. Our challenge is to apply naval power through day-to-day competition in a way that protects vital national interests and preserves regional security without undermining trust and triggering conflict.

These challenges create a unique – but limited – window of opportunity to chart a new course for American naval power in the Arctic Region. A Blue Arctic requires a new approach by the Navy-Marine Corps team to modernize the future naval force to preserve our advantage at sea and advance U.S. interests in the region.

To do so, we will build on our long history of presence and partnerships in the Arctic Region. Over 150 years ago, *USS Jamestown* stood our northern watch as the U.S. flag was raised over Alaska. Since then, our Sailors and submarines were the first to reach the North Pole, departing from our shores and those of our allies and partners. Our Marines have long trained and operated in the Arctic. During the Aleutian campaign in World War II, our naval forces bravely fought alongside our joint and allied partners to repel the enemy's attack. It was the proficiency and forward presence of American naval power in the Arctic Region that helped bring a peaceful end to the Cold War.

This regional blueprint is guided by the objectives articulated in the *National Security Strategy, National Defense Strategy, Department of Defense Arctic Strategy,* and *Advantage at Sea: Prevailing in Integrated All-Domain Naval Power;* supported by the *U.S. Navy Strategic Outlook for the Arctic* and informed by the *U.S. Coast Guard's Arctic Strategic Outlook.* Our naval forces will operate across the full range of military missions to deter aggression and discourage malign behavior; ensure strategic access and freedom of the seas; strengthen existing and emerging alliances and partnerships; and defend the United States from attack.

Naval forces will preserve peace and build confidence among nations through collective deterrence and security efforts that focus on common threats and mutual interests in a Blue Arctic. This requires an unprecedented level of critical thinking, planning, integration, and interoperability among our joint forces and international partners, along with greater cooperation among U.S. interagency, state, local, and indigenous communities.

In the decades ahead, the Department will maintain enhanced presence, strengthen cooperative partnerships, and adapt our naval forces for a Blue Arctic. We will work closely with partners – especially the U.S. Coast Guard, while building new partnerships, particularly in our Alaskan Arctic and the shores of our northern states. In doing so, we will provide our Sailors, Marines, and Civilians with the education, training, and equipment necessary to preserve peace and respond to crises in the region.

The crew of the Seawolf-class fast-attack submarine USS Connecticut (SSN 22) enjoys ice liberty after surfacing in the Arctic Circle during Ice Exercise (ICEX) 2020. (U.S. Navy photo by Mass Communication Specialist 1st Class Michael B. Zingaro)

The United States will always seek peace in the Arctic. History, however, demonstrates that peace comes through strength. In this new era, the Navy-Marine Corps team, steadfast with our joint forces, interagency teammates, allies and partners, will be that strength.

# [ CHALLENGES IN A NEW ERA ]

*Melting sea ice is making Arctic waters more accessible and navigable, enabling greater trade in the coming decades. Our Alaskan coast is already witnessing increased traffic.*

The coming decades will witness significant changes to the Arctic Region. Encompassing about six percent of the global surface, a Blue Arctic will have a disproportionate impact on the global economy given its abundance of natural resources and strategic location. The region holds an estimated 30% of the world's undiscovered natural gas reserves, 13% percent of global conventional oil reserves, and one trillion dollars' worth of rare earth minerals. Of the oil and gas reserves present in the Arctic, an estimated 84% likely reside offshore. Fish stocks are expected to continue to shift northward, attracting global fishing fleets and creating potential challenges to the current international prohibition on Arctic fishing.

Melting sea ice is making Arctic waters more accessible and navigable, enabling greater trade in the coming decades. Our Alaskan coast is already witnessing increased traffic. Russia is developing the Northern Sea Route to enable greater transit of military and commercial vessels alike. Canada has reinvigorated commercial activity along the Northwest Passage. Shipping traffic is rising with increased regional demand and movement of natural resources to markets, but will remain constrained by weather uncertainties, draft limitations, and costs. Port infrastructure is being developed to support maritime activity and local communities as ice recedes.

An F-22 Raptor with the North American Aerospace Defense Command intercepts a Russian Tu-142 maritime reconnaissance aircraft entering the Alaskan Air Defense Identification Zone. (Courtesy photo)

While only a fraction of global maritime activity transits the Arctic today, commercial activity is increasing through key strategic chokepoints such as the Bering Strait, Bear Gap, and Greenland-Iceland-United Kingdom (GIUK) Gap. The projected opening of a deep-draft trans-polar route in the next 20-30 years has the potential to transform the global transport system.

New commercial technologies will increase access and change the character of competition in the Arctic. Technological advancements will aid exploration and extraction of natural resources as well as the development of infrastructure and communications. Increased scientific expeditions will yield dual-use understanding of the maritime environment – and a potential military advantage. Our local Alaskan and indigenous communities will increasingly depend on the sea for trade and transport. Degrading permafrost puts Arctic infrastructure at risk. Arctic marine tourism will rise. Together, these changes will affect the fragile ecosystem and human safety, given the vast distances, harsh weather, and limited emergency response capabilities.

Greater access is further opening new Arctic undersea fiber optic cable routes linking Europe, Asia, and North America. Nation-states and other actors are mounting cyberattacks on Arctic ship-building, energy, and shipping sectors–especially the research and development communities that underpin them.

A U.S. Marine participates in cold-weather training and mountain-warfare training in Sorreisa, Norway. (U.S. Marine Corps photo by Cpl. Menelik Collins)

Rising maritime activity is spurring Arctic states to posture their navies to protect sovereignty and national interests while enabling their ability to project power. Nations in and out of the region are making investments in security and defense to enable Arctic operations. Arctic States – especially Russia – are reopening old bases, moving forces, and reinvigorating regional exercises. These trends will persist in the decades ahead.

Russia is investing heavily to enhance its Arctic defense and economic sectors, with a resultant multilayered militarization of its northern flank. By modernizing its military capabilities and posture – particularly the Northern Fleet – Russia aims to improve command and control, infrastructure, and joint force employment to project power and defend its northern approaches. In doing so, the escalatory and non-transparent nature of Russia's military activity and unlawful regulation of maritime traffic along the Northern Sea Route undermines global interests, promotes instability, and ultimately degrades security in the region.

The People's Republic of China views the Arctic Region as a critical link in its One Belt One Road initiative. As witnessed in other regions, a combination of Chinese capital, technology, and experience has the potential to influence Arctic shipping routes and undermine the economic and social progress of peoples and nations along these routes. China is investing in ship building – polar-capable cargo vessels, liquefied natural gas tankers, and nuclear-powered icebreakers – as well as port infrastructure to improve access in the Arctic. China's investments, global fishing fleet, and scientific, economic, and academic linkages to the people and institutions of Arctic nations, including joint ventures with Russia, will likely continue to rise in the decades ahead. We also expect increased Chinese Navy deployments

Naval forces will continue to maintain a routine presence on, under and above Arctic waters. (U.S. Navy photo by Mass Communication Specialist 2nd Class Jonathan Nelson)

on, below, and above Arctic waters. China's growing economic, scientific, and military reach, along with its demonstrated intent to gain access and influence over Arctic States, control key maritime ports, and remake the international rules-based order presents a threat to people and nations, including those who call the Arctic Region home.

Crewmembers aboard the Coast Guard Cutter Healy (WAGB 20) deploy a Sea Glider unmanned underwater vehicle to support scientific research in the Arctic Ocean. (U.S. Coast Guard photo by Petty Officer 3rd Class Lauren Steenson)

*A changing Arctic Region increases the potential for competition and conflict.* An opening Arctic presents a new operating environment for naval forces – an open maritime domain.  No longer limited to air, undersea, and strategic strike capabilities, rapidly melting sea ice increases Arctic access for surface vessels – both manned and unmanned.

The effects of new technologies, economic drivers, and competition among major powers are influencing regional security and stability. These conditions require a naval force to match this reality. Yet no one nation has the knowledge and resources required to provide security and defense throughout the entire Arctic Region. For that reason, we seek to cooperate with Allies and partners in the Arctic.

The Department of the Navy – working closely with our joint force, the interagency, allies, and partners – will maintain an enhanced presence, strengthen cooperative partnerships, and build more capable Arctic forces to achieve common interests and address these challenges in the coming decades.

*The Navy's Arctic resilience is embodied in Rear Admiral Peary's famous motto: "I will find a way or make one!"*

# [ REGIONAL BLUEPRINT ]

*The Department will maintain an enhanced presence in the Arctic Region by regionally posturing our forces, conducting exercises and operations, integrating Navy-Marine Corps-Coast Guard capabilities, and synchronizing our Fleets.*

An increasingly accessible and navigable Arctic operating environment will place new demands on our naval forces. The scope and pace of our competitors' and adversaries' ambitions and capabilities in a Blue Arctic requires new ways of applying naval power. The Arctic Region is a vast maneuver space and this regional blueprint recognizes the rising importance of enhanced naval presence and partnerships in the region. Flexible, scalable, and agile naval forces provide an inherent advantage in a Blue Arctic, but it is necessary to enhance our presence, cooperation, and capabilities. Concurrently, we will find new ways to integrate and apply naval power with existing forces while investing in new capabilities that may not be fully realized and integrated into the force for at least a decade.

We will achieve our enduring national security interests in a Blue Arctic by pursuing these objectives:

» Maintain Enhanced Presence;

» Strengthen Cooperative Partnerships; and

» Build a More Capable Arctic Naval Force.

The Seawolf-class fast attack submarine USS Seawolf (SSN 21) conducts operations in the Norwegian Sea off the coast of Tromsø, Norway. (U.S. Navy courtesy photo)

## Maintain Enhanced Presence

This regional blueprint underscores the use of naval power to influence actions and events at sea and ashore. Left uncontested, incremental gains from increased aggression and malign activities could result in a fait accompli, with long-term strategic benefits for our competitors. The U.S. Navy currently has routine presence on, under, and above Arctic waters, and we will continue to train and exercise to maximize this capability. *The Department will maintain an enhanced presence in the Arctic Region by regionally posturing our forces, conducting exercises and operations, integrating Navy-Marine Corps-Coast Guard capabilities, and synchronizing our Fleets.*

**Regionally postured naval forces.** Our security and prosperity are inextricably linked with other nations. Without adequate defensive posturing, competition over Arctic resources and sea routes could present a direct threat to U.S. sovereignty. The Department will be postured to deter aggressive and malign behavior, keep the seas free and open, and assure allies and partners of our long-term commitment to preserving peace and advancing shared interests.

In the decades ahead, the Department will continue to provide the right levels and types of presence overseas. We will continue to assess our force posture requirements in Alaska to meet the unique and evolving requirements of the Arctic Region, often in conjunction with the joint force, U.S. interagency, allies, and partners – including NATO. The Department will take a more cooperative and tailored approach through a mix of permanently stationed forces, rotational forces, temporary forces, pre-positioned equipment and stocks, and basing infrastructure across the region.

A U.S. Marine provides security at Fort Greely, Alaska, in preparation for exercise Arctic Edge 2020. (U.S. Marine Corps photo by Lance Cpl. Jose Gonzalez)

Credible naval forces ensure the ability to deter competitors and rapidly respond to crises in the region, while allowing naval forces to project power to gain an advantage in other theaters. In addition to our own forces, effective Theater Security Cooperation activities are a form of extended deterrence, creating security and reducing conditions for conflict. To this end, the Navy-Marine Corps team will conduct exercises and operations, key engagements, and port calls with allies and partners across the Arctic Region. Our defense posture must be regularly and rigorously assessed to adapt to a Blue Arctic.

**Exercises and operations.** U.S. naval forces participate in a wide spectrum of regional exercises and we will continue to build off this foundation. The submarine-focused Ice Exercise (ICEX) has been held in the Beaufort Sea and Arctic Ocean since the 1960s, making it the longest running Arctic exercise. NATO Exercise Trident Juncture marked the largest post-Cold War exercise in the region, bringing together nearly 50,000 personnel, 65 ships, and 250 aircraft for a collective defense exercise in Norway and surrounding waters. We integrated naval forces and cold-weather amphibious operations during the Arctic Expeditionary Capabilities Exercise in Alaska. During Operation Northern Edge, our Department exercised joint operations with 10,000 personnel from the USS Theodore Roosevelt Carrier Strike Group, Air Force, Army, and Marines – all united in efforts to enhance interoperability and improve operational capabilities.

The Royal Canadian Ship MV Asterix conducts a replenishment-at-sea with the guided-missile destroyer USS Thomas Hudner (DDG 116) and Royal Canadian Navy frigate HMCS Ville de Québec during Operation Nanook. (U.S. Navy photo by Mass Communication Specialist 2nd Class Sara Eshleman)

The U.S. Navy routinely exercises with allies and partners in the Arctic Region. Exercise Dynamic Mongoose brings together NATO surface ships, submarines, and maritime patrol aircraft for complex anti-submarine warfare training off the coast of Iceland. Operation Nanook-Nunalivut tests U.S., Canadian, French, and Danish warships in the North Atlantic above the Arctic Circle. Marines hone their cold-weather fighting skills through exercises such as Cold Response and Arctic Edge.

The U.S. Second and Sixth Fleets routinely conduct operations in the Arctic Region, demonstrating unique capabilities such as operating an expeditionary maritime operations center from Iceland, multi-ship surface action groups patrolling the Barents Sea, and vigilant P-8A aircraft conducting anti-submarine operations in strategic areas. Our submarines will continue to operate beneath the waves, even as ice diminishes.

In the years ahead, U.S. naval forces will continue to deploy in the Arctic Region. We will increase our participation in regional exercises and operations across all Arctic sub-regions, demonstrating U.S. commitment to the region – and to our allies and partners. This includes working with our Arctic experts in Alaska, such as the Alaskan State Defense Force and Alaska National Guard. Naval forces will enhance operational proficiencies by participating in joint, bilateral, and multilateral exercises to improve interoperability, warfighting prowess, and operational expertise in a Blue Arctic.

Naval forces will fly, sail, and operate to preserve freedom of the seas, defend the homeland, and act as a credible deterrent. We will ensure common domains remain free and open, and prevent competitors from disrupting or controlling

Arctic sea lines of communication and commerce. When disasters and accidents strike, the Department will be ready to operate with the Coast Guard, U.S. interagency, and our international partners. As we meet the demands of a Blue Arctic, the Department remains committed to protecting the Arctic environment and ensuring naval forces do their part to help assess and preserve it.

**Integrated naval power in Arctic littorals.** The vast, remote, and austere Arctic environment provides naval forces an opportunity to exploit key terrain to improve the security of sea lines of communications and choke-points while enhancing the natural barriers formed by Arctic littorals. Operating in these difficult conditions will challenge naval forces in ways not experienced in generations. A unified approach is essential.

This regional blueprint challenges the Navy-Marine Corps-Coast Guard team to evolve an expanded range of integrated capabilities to achieve enduring national interests in the Arctic Region. We will organize, train, and equip as a naval expeditionary force capable of operating in Arctic littorals. The Marine Corps will facilitate sea control and sea denial operations in support of Fleet commander plans. Integrating the Coast Guard's unique authorities and capabilities with the Navy-Marine Corps team expands options for Fleet Commanders in day-to-day operations and crisis. This combination of resilient, mobile, and self-sustaining forces will enable us to persist, partner, and operate in the Arctic Region.

Flags from partner nations fly over Ice Camp Seadragon during Ice Exercise (ICEX) 2020. ICEX is a biennial exercise which promotes interoperability between allies and partners to maintain operational readiness and regional stability. (U.S. Navy photo by Mass Communication Specialist 1st Class Michael B. Zingaro)

**Fleet synchronization.** Our forces must be unified in their preparedness to maintain sea control and project naval power in, from, and to the Arctic Region. To do so, we will further enhance Fleet synchronization across Combatant Commands by improving command relationships, coordination, and connectivity.

## Strengthen Cooperative Partnerships

Mutually beneficial alliances and partnerships, are foundational to this regional blueprint. Competitors seeking to disrupt the international rules-based order in the Arctic must be met with a firm commitment of like-minded naval forces and nations to address shared challenges and uphold regional interests and responsibilities. When we pool resources, leverage our comparative advantage, and share responsibility for our common defense, our collective security burden becomes lighter. We will cooperatively identify ways to generate synergies from each other's postures and capabilities to confront shared regional threats. Allied and partner naval forces must jointly assess threats, define roles and missions, deepen defense industrial cooperation, and develop and exercise new concepts of operations for the Arctic Region. Equitable burden sharing is necessary and will take time, but the process of doing so will strengthen our collective capabilities.

The Military Sealift Command fast combat support ship USNS Arctic (T-AOE-8) prepares to conduct a replenishment-at-sea while operating alongside the aircraft carrier USS Harry S. Truman (CVN 75). (U.S. Navy photo by Mass Communication Specialist 3rd Class Joshua Moore)

We will strengthen existing partnerships and attract new partners to meet shared challenges, opportunities, and responsibilities in the Arctic. Together we will enhance our awareness, expand collaborative planning, and improve interoperability. In doing so, enhanced and predictable cooperative activities enable naval forces to maintain credible presence and deter malicious activity. Naval forces are stronger when we operate jointly and together with allies and partners.

**Enhance awareness.** To be effective, we must clearly define and prioritize threats and vulnerabilities. As such, naval forces must gather, synthesize, and share information and intelligence across the joint force and U.S. interagency, and with allies and partners. We must seek new and innovative approaches to connect people and information to help us avoid tactical and strategic surprise. To this end, understanding and predicting the physical environment from sea floor to space, today and for decades, is critical for mission advantage, ensuring the safety of personnel and equipment, and informing future force requirements.

**Expand regional consultative mechanisms and collaborative planning.**
Communication, constraint, transparency, and verification are foundational to preventing unintended military escalation in the Arctic. In the
decades ahead, the Department will expand participation in regional
consultative mechanisms and collaborative planning for a peaceful Blue
Arctic. This includes reinforcing existing mechanisms, such as the bilateral Prevention of Incidents On and Over the High Seas (INCSEA), and the
United Nations Convention for the Law of the Sea (UNCLOS), to reduce the
potential for misperceptions, accidents, and unintended conflict among
forces operating in the Arctic.

U.S. Marines conduct a foot patrol during cold-weather training in Bjerkvic, Norway, in preparation for
Exercise Cold Response designed to enhance military capabilities and allied cooperation in a challenging
arctic environment. (U.S. Marine Corps photo by Cpl. Isaiah Campbell)

**Improve interoperability and collaboration.**  The Department will increase
interoperability with the joint force, interagency, allies, and partners as we
seek to boost Arctic capabilities.

»  *Allies and Partners.* We have a proud tradition of working together
with partners in the Arctic Region – Lieutenant Commander
Robert Byrd departed from Svalbard, Norway on his 1926 flight
to the North Pole. We must enhance existing and emerging partnerships in ways that address priority missions, while ensuring
mutually beneficial partnerships. In doing so, we will improve
interoperability to ensure our forces can operate seamlessly
across the full spectrum of presence and warfighting operations
in day-to-day competition and crises.

We will prioritize security cooperation efforts to support our allies and partners as they seek to improve their capabilities, while increasing our participation in regional exercises – in terms of the quantity of forces, number of exercises, and types of exercises – to demonstrate our commitment and enhance our collective capabilities. Given competing demands on our forces, this will require innovative approaches that integrate new and existing platforms and people to broaden Arctic exposure across our naval force.

» *New Partners.* All Arctic nations have a vested interest ensuring the region remains safe and stable. The U.S. Navy has an impressive history of enhancing and deepening coordination and cooperation with all Arctic nations in a number of areas – from annual staff talks and Incidents at Sea Treaty (INCSEA) meetings to conducting ship visits, naval games, and combined exercises. We will do so where mutually beneficial.

» *Joint Force.* The Navy-Marine Corps team will work with the joint force to synchronize our plans, integrate our capabilities, and gain broader maritime domain awareness and C5ISR in the Arctic. We will strengthen coordination and interoperability with the U.S. Coast Guard – especially their indisputable expertise for ice-breaking missions.

» *U.S. Interagency.* Naval forces will work in concert with interagency efforts to create opportunities for dialogue, coordination, and mutually beneficial partnerships. In doing so, we will maximize transparency and ensure our planning efforts are integrated with local, state, federal, and indigenous communities. Native Alaskan tribes have a long and honorable history of military service that blend seamlessly with such efforts.

» *Public-Private Partnerships.* Public-private partnerships and innovative solutions, such as NavalX tech bridges, leveraging relevant civilian expertise within the reserve components, and the Alaska Regional Collaboration for Technology Innovation and Commercialization (ARCTIC) led by the Office of Naval Research will help transform the operating abilities – and regional understanding – of naval forces. Maine has taken bold steps to advance their Arctic connection, including recent participation in The Arctic Circle Assembly and relationship with Eimskip, an Icelandic international shipping company, which has a regional hub in Portland. *The governor of Maine recently stated "Maine is poised to become the hub, the eastern gateway to the Arctic, a region whose allure we have shared since Portland, Maine explorer Robert Peary set foot on the North Pole in 1909."*

A U.S. Marine Corps Mountain Warfare Training Center chief instructor places a USMC flag in the ice outside of Ice Camp Seadragon during Ice Exercise (ICEX) 2020. (U.S. Navy photo by Mass Communication Specialist 1st Class Michael B. Zingaro)

## Build a More Capable Arctic Naval Force

Following the Cold War, the Navy-Marine Corps capabilities and operational expertise in the Arctic diminished. Recent efforts to increase our capabilities have improved operational readiness, which is required regardless of ice conditions and time of year. Though we routinely patrol on, above, and below Arctic waters, the Department must be prepared and postured to meet the demands of an increasingly accessible Arctic operating environment.

**Modernize capabilities.** We will invest in key capabilities that enable naval forces to maintain enhanced presence and partnerships. While meeting global requirements, we will take a targeted, disciplined, and coordinated approach to balancing our personnel, platforms, and posture to solidify our competitive advantage in the Arctic Region. In doing so, the Department will work closely with the Office of the Secretary of Defense, Joint Staff, and Combatant Commanders to identify and meet requirements for our defense planning scenarios, campaign plans, and deployment models.

The Royal Navy frigate HMS Kent (F78), the guided-missile destroyers: USS Roosevelt (DDG 80), USS Porter (DDG 80), USS Donald Cook (DDG 75), and the fast combat support ship USNS Supply (T-AOE-6) conduct joint maritime security operations in the Arctic Ocean. (Courtesy photo)

The Department will continue leading critical advancements in research, development, testing, and evaluation – including the development of cold weather-capable designs, forecasting models, sensors, high latitude communications, and navigation systems – while enhancing our ability to meet future demands.

> » *Infrastructure.* Access to infrastructure – port facilities, airfields, and shore infrastructure – across the Arctic is critical for naval forces to project power. To maintain our operational advantage in a Blue Arctic, we must explore opportunities to reduce transit times, preserve mobility, and meet logistical demands of naval forces operating throughout the Arctic Region. The Department of the Navy will ensure investments in installations match

future operational needs. This means working closely with our joint and interagency partners – as well as regional allies and partners – to meet the strategic demands of a Blue Arctic.

» *Command, control, communications, computers, cyber, intelligence, surveillance, and reconnaissance (C5ISR).* The Department will assess and prioritize C5ISR capabilities in the Arctic Region, to include resilient, survivable, and interoperable networks and information systems of naval tactical forces, operations centers, and strategic planning. These capabilities will enhance domain awareness with the joint force, U.S. interagency, allies, and partners in the Arctic Region.

» *Naval Forces.* The Department will evaluate and modernize existing and future forces to provide manned and unmanned operational presence and patrol options in cold weather and ice-diminished Arctic waters. We will improve hydrographic surveys and sensors to support the Fleet. In a Blue Arctic, the Department must have a more credible presence in Arctic waters. This means ensuring that Arctic operations are considered in our design and modernization plans, and that our defense industrial base can build and sustain forces for the Arctic.

» *Science and Technology.* Progress is not possible without pushing the boundaries of science and technology. The Office of Naval Research (ONR) will continue efforts to enable naval presence and partnerships in basic and applied research across the spectrum of the physical and social sciences. We will continue to spearhead programs like the Arctic Mobile Observing System (AMOS) Innovative Naval Prototype, lead

The Coast Guard Cutter Stratton (WMSL 752) patrols above the Arctic Circle near the Bering Strait in support of Operation Arctic Shield. (U.S. Coast Guard photo by Lt. Brian Dykens)

the International Cooperative Engagement Program for Polar Research (ICE-PPR) and conduct research at Naval Research Laboratories, warfare centers, and academic institutions, like the Naval Postgraduate School and U.S. Naval Academy, while collaborating with other federal and international partners to enhance capabilities and trusted partnerships.

An Air-Deployable Expendable Ice Buoy is deployed in the Arctic from a Royal Danish Air Force C-130 as part of the International Arctic Buoy Program. (U.S. Navy photo by John F. Williams)

**Evolve innovative operational concepts.** Modernizing our naval forces for a Blue Arctic goes beyond just facilities and platforms. It requires adjusting how we organize and employ naval forces. We must anticipate the challenges of a Blue Arctic and develop and test operational concepts to sharpen our advantage at sea. This will include concepts for Distributed Maritime Operations (DMO), Littoral Operations in Contested Environment (LOCE), and Expeditionary Advanced Base Operations (EABO).

**Prepare our people.** Sailors, Marines, and Civilians are the greatest enduring strength of our Department. Our principal emphasis must focus on developing, educating, and training our people for a Blue Arctic.

» *Professional Military Education (PME).* PME on the Arctic Region is limited in scope and quantity. We must strengthen cooperation across PME institutions and expand course offerings – in resident and distant learning – that deepen our knowledge of current and future challenges, historical Arctic operations, and the strategies and tactics to counter competitors.

» *Training.* We will develop new ways to improve integration between education and training commands by including curricula on Arctic operations and injecting emerging naval concepts into training pipelines to help our Sailors, Marines, and Civilians gain and share the skills required to operate in the region. Critical to enhancing Arctic partnerships is the development of cultural and operational expertise and appropriate detailing to impactful, career enhancing billets. The Department will emphasize cold weather training to provide modern procedures and guidance for Arctic operations.

» *Personnel Exchange Programs.* We will increase personnel exchange and interoperability programs with the U.S. Coast Guard as well as allied and partner forces in the region to enhance our understanding of the Arctic and improve cooperation. This includes professional military education institutions, operational units, and research, development, testing, and evaluation (RDT&E) communities.

**Implementation.** This regional blueprint presents a sea change for naval forces in the Arctic Region. Our renewed and strategic efforts will improve the readiness, capabilities, and professionalism of naval forces in the coming decades. In doing so, the Department must have a unified, deliberate, and forward-looking approach to regional investment in infrastructure and force development.

A U.S. Navy explosive ordnance disposal (EOD) technician and a Royal Norwegian Navy EOD commando place a training mine for a follow-on cold water advanced neutralization procedure during exercise Arctic Specialist held in Ramsund, Norway. (U.S. Navy courtesy photo)

A U.S. Navy logistics specialist from the Arctic Submarine Laboratory stands outside Camp Seadragon during Ice Exercise (ICEX) 2020. (U.S. Navy photo by Mass Communication Specialist 1st Class Michael B. Zingaro)

*U.S. naval forces have long held cooperative partnerships with Arctic States and like-minded non-Arctic States with regional interests. We will continue to strengthen these partnerships while attracting new partners to meet the shared challenges, opportunities, and responsibilities of a Blue Arctic.*

U.S. service members assigned to Underwater Construction Team One (UCT-1) and military personnel from Norway and the Netherlands pose for a photo on a frozen lake in Skjold, Norway, during Exercise Cold Response 2020. (U.S. Navy photo by Mass Communication Specialist 2nd Class Mark Andrew Hays)

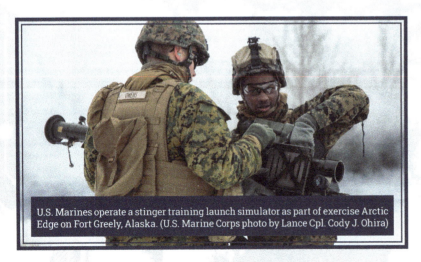

U.S. Marines operate a stinger training launch simulator as part of exercise Arctic Edge on Fort Greely, Alaska. (U.S. Marine Corps photo by Lance Cpl. Cody J. Ohira)

A U.S. Navy Diver assigned to Mobile Diving and Salvage Unit 2, performs an ice dive in the Arctic Circle during Ice Exercise (ICEX) 2016. (U.S. Navy photo by Mass Communication Specialist 2nd Class Tyler N. Thompson)

U.S. Marines conduct multilateral training alongside the Netherlands Royal Marines, German Sea Battalion and Belgian Paratroopers during the Arctic Movement and Survival Course in Norway. (U.S. Marine Corps courtesy photo)

# [ CONCLUSION ]

*The Navy-Marine Corps team, steadfast with our joint forces, interagency teammates, allies, and partners, will maintain an enhanced naval presence, strengthen cooperative partnerships, and build more capable naval forces for a Blue Arctic.*

The Department of the Navy has a proud history of operating in the Arctic. American ingenuity has continuously enabled U.S. naval forces to prevail in the region. The Navy's Arctic resilience is embodied in U. S. Navy Rear Admiral Peary's famous motto: *"I will find a way or make one!"* Just as Peary, a Civil Engineer Corps Officer, led successful Arctic expeditions – together with Matthew Henson, Ootah, Egigingwah, Seegloo, and Ooqueah – this regional blueprint recognizes the long-term challenges and opportunities of a Blue Arctic – and the role of American naval power in it.

This regional blueprint reflects America's key security interests and objectives in the Arctic Region. We seek peace, cooperation, and stability, but history has proven that peace is enabled by strength. The Navy-Marine Corps team, steadfast with our joint forces, interagency teammates, allies, and partners, will maintain an enhanced naval presence, strengthen cooperative partnerships, and build more capable naval forces for a Blue Arctic. Following these imperatives will ensure we remain the most ready, respected, and capable naval force in the world, which is what our nation expects and deserves.

Rear Admiral Lorin Selby, chief of naval research, signs the International Cooperative Engagement Program for Polar Research (ICE-PPR) memorandum of understanding as the U.S. executive member for the Office of the Secretary of Defense. (U.S. Navy photo by John F. Williams)

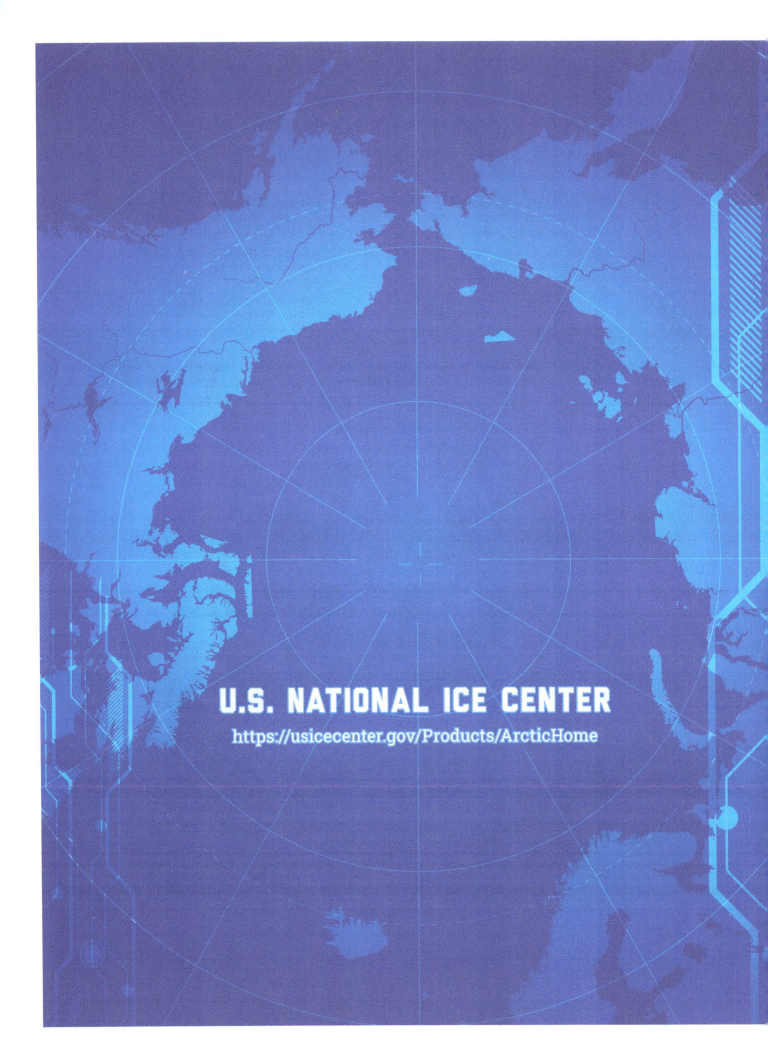

U.S. NATIONAL ICE CENTER

https://usicecenter.gov/Products/ArcticHome

www.ingramcontent.com/pod-product-compliance
Lightning Source LLC
Chambersburg PA
CBHW080545060326
40690CB00022B/5227